I0605346

IN-DEMAND CAREERS

BE AN

INFORMATION SECURITY ANALYST

by Miles Herman

BrightPoint Press

San Diego, CA

an imprint of ReferencePoint Press, Inc.
Printed in the United States

For more information, contact:
BrightPoint Press
PO Box 27779
San Diego, CA 92198
www.BrightPointPress.com

LIBRARY OF CONGRESS CATALOGING-IN-PUBLICATION DATA

Name: Herman, Miles, author.
Title: Be an information security analyst / by Miles Herman.
Description: San Diego, CA: ReferencePoint Press, 2026 | Series: In-demand careers | Audience: Grade 7 to 9 | Includes bibliographical references and index.
Identifiers: ISBN: 9781678211189 (hardcover) | ISBN: 9781678211196 (eBook)
The complete Library of Congress record is available at www.loc.gov.

CONTENTS

AT A GLANCE

- Hackers break into computer systems. They often steal data or damage the systems by leaving viruses.
- Information security analysts, also called infosec analysts, stop hackers. They protect data and computer systems.
- Infosec analysts use security software in their jobs. This software spots threats and sends alerts.
- There are several ways to become an infosec analyst. These include bootcamps, colleges, and certification programs. Free training programs are available, too.
- Students can prepare in high school for this career. They can take computer science classes.
- Some infosec analysts focus on certain areas. They are called specialists.

- The job market for infosec analysts is growing quickly. Employers want to hire these roles to protect important information.

- In 2024, the average salary for an entry-level analyst was $103,000 per year.

FROM GAMER TO INFORMATION SECURITY ANALYST

Anthony Gomez had always loved video games. He played his favorite game a lot. He spent his free time trying to move on to different levels. He focused on earning power perks and boosts. These helped him advance quickly through the game. But it was hard work. It also took up a lot of time.

One day, Gomez had an idea. He wondered if he could hack into the video game. Then he might be able to get

Cheat codes in video games change how a game works. They can make gameplay easier or harder.

more boosts. He searched the internet and learned more about computer programming. He found an online community of others like him. Gomez learned everything he could. He studied the techniques hackers used. In high school, he took a college-level anti-hacking class. The teacher worked in cybersecurity. He told Gomez he could turn his interest in hacking into a career.

INFORMATION SECURITY

Gomez decided to become an information security analyst. They are also called infosec analysts. Some infosec analysts go to college to study. But Gomez did not. He entered a free training program after high school. He studied cybersecurity

Hackers can cause computers to have errors.

for 6 months. He learned about how computers worked. He also learned about safely storing data. Gomez learned the skills he needed to be an analyst. Then he got an internship with a large company called Hulu. An internship is a short-term job. It is a way to learn more about a career. An internship does not guarantee a permanent job. But Hulu hired Gomez after his internship ended.

Internships allow people to observe workers and ask questions about day-to-day tasks.

Now, Gomez is an infosec analyst. Gomez's job is to protect computer systems and their data. A computer system includes hardware and software. Hardware refers to the physical parts of a computer. These include screens, printers, and **hard drives**. Software refers to programs that make computers perform certain tasks. Gomez's job is to look out for hackers and threats.

Hackers break into computer systems. They steal data or plant viruses. A virus is a program that can damage computers.

Gomez enjoys his work, "It's never the same day for me," he says. "[My job] allowed me to learn a lot of valuable skills that I can now pass on to anyone else trying to make it up through a nontraditional way."[1]

Infosec analysts may study data using charts and graphs.

CHAPTER ONE

WHAT DOES AN INFORMATION SECURITY ANALYST DO?

Infosec analysts do important work. They protect and monitor computer systems and data. They stop hackers and help get rid of viruses. Hackers are always at work. They try to break into computer systems. Hackers steal important information such as passwords. Many companies store lots of personal information. Hackers may target them. Sometimes a company must pay to get back stolen data. If news gets out,

Hacker attacks can cost people millions of dollars.

the company may lose customers. People will not trust the company to keep their data safe. The company needs the help of infosec analysts. They help update the company's security systems and software. Updates helps prevent future harm.

An infosec analyst may also educate users on how to keep their computers safe. For example, analysts warn users about

The Largest Data Theft in History

In 2013, hackers broke into Yahoo's computer system. Yahoo provides internet services. Hackers stole the data of 3 billion users. They took names, addresses, and phone numbers. Many customers stopped using Yahoo after the theft. They were angry with the company for not securing their data. This was the largest theft of data in history.

It is important to double-check the sender of an email before clicking on a link. A bad link can lead to viruses.

different viruses. They teach users how to spot bad emails. They let users know when to update passwords. Users may also contact infosec analysts with questions. Some users have security problems. Others need help protecting their data.

TYPICAL DUTIES

Infosec analysts ensure computer systems are safe. They make sure everything is

Servers are computers that provide service and data to other computers. Infosec analysts help protect computer servers.

working properly. This means they watch closely for anything unusual. Programs may slow down or freeze. Data could vanish. **Pop-up windows** may show up on screens. These could be signs of **malware**.

Infosec analysts use software to help with tasks. Security software is a vital tool for protecting computers. **Firewalls** limit who can use a computer. They stop intruders from getting inside computer systems. Anti-virus programs spot strange patterns. They send alerts when a virus or threat is present.

Programs used by infosec analysts may send alerts when something is wrong. Infosec analysts look into these alerts. They decide whether a threat is serious. Then they figure out who is at risk. The threat could be small. But large attacks can do a lot of damage. Hackers can shut down entire systems. They can steal data and passwords. Some attacks hurt the public. In 2015, malware shut off a large portion of Ukraine's power. It left more than 230,000 people in the dark. This is why infosec analysts keep software up-to-date. They may install new software to fix problems.

Infosec analysts follow written steps to handle serious threats. They usually gather a team. It may include forensic investigators and incident managers. Forensic investigators collect evidence of a crime. Incident managers keep track of documents and make sure they are safe. The team reviews the threat and makes a plan to stop it from spreading. The team also tries to get back stolen data. Infosec analysts look for these solutions. They try to get computer systems up and running quickly. After the problem is solved, infosec analysts write a report. The report explains the threat. It also describes how the team responded and how to avoid future mistakes.

INFOSEC ANALYST SPECIALTIES

Infosec analysts may become specialists in their fields. Specialists focus on certain areas of cybersecurity. One area is auditing. A security auditor does a complete review of a computer system. They ensure companies are following information

Infosec analysts may work with information technology specialists to solve security threats.

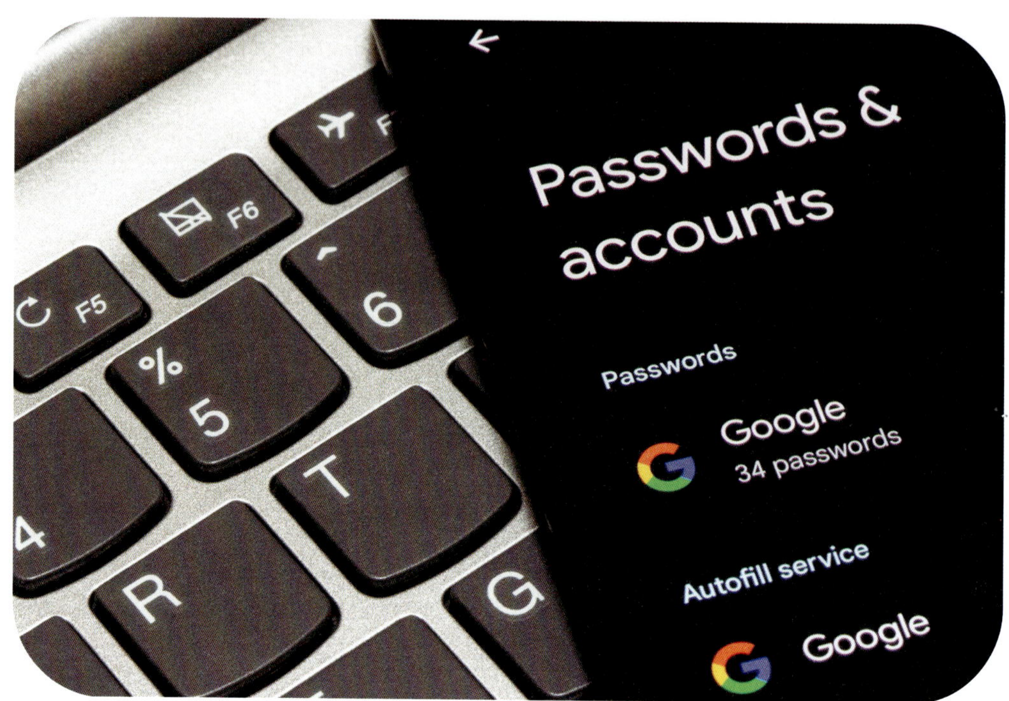

Some people use password managers to help keep track of complicated passwords.

technology safety laws. Auditors may also find out if users are not being safe. Users may be using old passwords that have been stolen. They may be clicking on bad links. This can allow viruses to steal private information.

Some specialists work as cryptographers. They use codes to protect data. The code is like a locked door.

Only people with a key can enter. One example of a code is a password. Users need passwords to get into certain computer programs. Most people use short passwords. Short passwords are easier to remember. But they are easier to guess. Cryptographers use tougher passwords. These passwords are usually much longer. They may contain numbers, letters, and unique symbols. Cryptographers also use technology programs to code data and messages. Encoded data and messages cannot be read by other people.

Ethical hackers are also specialists. Companies hire them to break into the companies' computer systems. Ethical hackers use the same hacking skills as those who cause harm. But they do

Infosec analysts must pay attention to details. They take note of ways a hacker can break in.

not misuse the information they access. Instead, they identify weaknesses. They work to keep hackers from abusing the same weaknesses in the systems. Jamie Woodruff is an ethical hacker. He used to be a black hat hacker. This is someone who hacks to cause harm. Now Woodruff uses his skills to help companies. He once was hired to test a bank's security system. He pretended to deliver pizza. The workers let him inside. Then he got into the bank's computer systems. "Hackers will adapt to any scenario that is put in front of them."[2]

Hacker attacks keep rising. This means more infosec analysts are needed. Around 17,000 new jobs for infosec analysts open every year. For those interested, there are many paths to becoming an infosec analyst.

CHAPTER TWO

BECOMING AN INFORMATION SECURITY ANALYST

There are many possible career paths for people who like technology. Information technology is one possible path. It requires skills beyond understanding computers.

Problem-solving is one of the most important skills. Infosec analysts study how hackers get into systems. They figure out if any data was stolen. They also look for malware. Then they try to fix any problems.

Most infosec analysts are part of a team. Infosec analysts must learn to communicate clearly with team members and users every day.

Taking computer science classes helps students learn how to write computer programs.

Infosec analyst Nick Boenzi says this is his favorite part of the job. He compares the work to solving a puzzle. "[I enjoy] figuring out what happened and how to prevent it from happening again," he says.[3]

HIGH SCHOOL COURSES

Students interested in infosec can take STEM classes in high school. STEM stands for science, technology, engineering, and math. Students may learn to write programs. Programming classes teach students how to develop software. Writing and speech classes are also helpful. Infosec analysts write reports. They talk to many people. Sometimes they train users. Strong communication skills can help infosec analysts.

Cybersecurity Competition

CyberPatriot is a US government program. It teaches youth about cybersecurity. It runs the largest cybersecurity competition for students in the world. In 2024, more than 5,000 teams competed. They looked for problems in a computer system. Then they tried to fix them. Only twenty-eight teams reached the finals. Winners received scholarships and other prizes.

BOOT CAMPS

Boot camps prepare students for specific jobs. These camps can be in person, online, or both. Many camps require students to have a high school or General Education Development (GED) diploma to enroll.

Infosec analyst boot camps are good for people looking for **entry-level** jobs. Students learn how to spot cyber threats and solve security issues. They work with security software. They test systems for weaknesses. Some boot camps teach more advanced subjects. These include forensics, cryptography, and ethical hacking. Most boot camps usually last up to 6 months. There is a lot of information to cover in a short time. Students may need to study up to 90 hours per week. This can be difficult for people with jobs.

Many boot camps have job counseling. Students can get help applying for work. Some boot camps also have job placement programs. Boot camps are often cheaper

Balancing too many responsibilities can cause stress and burnout.

than 4-year degree programs. However, some employers require a degree.

COLLEGE

College is another way to become an infosec analyst. Students can earn both associate's or bachelor's degrees. Associate's degrees are usually 2-year programs. Bachelor's degrees are usually 4-year programs.

Career website Roadtrip Nation reports that 13 percent of infosec analysts have associate's degrees. Many community colleges offer associate's degrees. These programs prepare students for entry-level jobs. They teach the basics. Two-year programs are longer than boot camps. The programs may also be less intense.

Study groups can be a helpful way to learn new information.

Students take other classes. Students can study subjects such as English and public speaking. This allows them to develop more well-rounded skills.

Roadtrip Nation reports more than half of infosec analysts have bachelor's degrees. Students in these programs learn cybersecurity basics. But they can study advanced subjects, too. Infosec analysts with bachelor's degrees often earn more than those with associate's degrees or boot camp training. They may have an easier time finding jobs. But 4-year programs generally cost more than boot camps and 2-year programs. Students will also need to

Graduation ceremonies are a way to celebrate students who have completed academic degrees.

take other classes to graduate. These could include English, science, and math.

CERTIFICATIONS

Certifications are official documents. They state that a person has the skills to do a certain job. They are not required. But once complete, a person gains additional skills. Certifications are offered in areas such as auditing, forensics, and ethical hacking for infosec analysts already in the field. Some certifications are also available for entry-level infosec analysts.

The process of getting certified is difficult. Infosec analysts must pass an exam in their chosen field of work. They need to study. They should have a good grasp of the topics and skills needed in order to pass.

US MILITARY

The US military has six branches. They are the Army, Navy, Air Force, Marines, Coast Guard, and Space Force. Each branch has a cybersecurity section. These sections protect military computer systems. Military systems have top-secret data. Some control weapons such as missiles. Others help during combat. Military analysts have critical jobs. They look out for hackers and threats.

The military trains people to become infosec analysts. **Recruits** can ask to join a cybersecurity section. They must take a test called the Armed Services Vocational Aptitude Battery (ASVAB). The test shows recruits' strengths and weaknesses. It suggests places to work. To become an infosec analyst, recruits have to meet a

Subjects covered on the ASVAB include general sciences, electronics information, and assembling objects.

minimum score. They must test well in the skills needed for the job.

The military pays for an analyst's education in exchange for years of service. Most military analysts make a lower salary. But they do not pay tuition. They also gain work experience.

FREE PROGRAMS

Some people take different career paths. Anthony Gomez entered the Year Up United program. This is a free training program for young adults who qualify. He learned basic skills. Then he completed an internship.

Some infosec analysts go to school after they start their jobs. That is what Gomez did. He says, "I'm also currently working toward my associate degree in information security to bolster my skills."[4] Companies may offer to pay tuition for their employees.

There are many pathways to an infosec career. The best path is different for each person. No matter what, people can get the training they need to succeed as an infosec analyst.

Infosec analysts work together to learn about new technologies.

CHAPTER THREE

A DAY IN THE LIFE OF AN INFORMATION SECURITY ANALYST

It can be hard to describe a typical day for an infosec analyst. This is because jobs look different at different companies. Casey Lems has worked in computer security for many years. He says, "Every company/job is different, so even if you get an idea from one job, only about 15 percent transfers to the next!"[5]

Around 75 percent of infosec analysts work for large employers. These employers

Infosec analysts may work in office buildings.

Some infosec analysts may travel to clients and businesses to do work.

have more than 1,000 employees. Infosec analysts are part of security teams. Some are specialists. They focus on areas such as audits and ethical hacking.

Around 10 percent of infosec analysts work for small employers. A small employer has fewer than 100 employees. These infosec analysts may work alone or with small teams. In a smaller company, infosec analysts often have more duties.

ENTRY-LEVEL ANALYSTS

Many entry-level analysts start with basic tasks. They monitor computer systems and handle threats. They train users and answer questions.

Entry-level analysts may start the day by checking their calendars and emails. They make sure they are on track for any deadlines. They also look for alerts

Infosec Analyst Consultant

Infosec analysts might work as consultants. Consultants are not usually permanent employees. Instead, a company hires a consultant for a set period of time to help with a problem. Some consultants work alone. Others work for consulting companies. When a company needs cybersecurity help, it can hire an infosec analyst as a consultant.

about hackers that may have popped up overnight. If there are any threats, they figure out whether they are serious. If there are no alerts, analysts may complete other tasks. They can research recent attacks at other companies. They can study new programming tools that stop hackers. This helps them prepare for any future threats.

Mid-morning tasks may include meetings. Entry-level analysts gather with team members. They talk about new threats. They follow up on old ones. The team leader may assign new tasks.

Entry-level analysts continue working on tasks for the rest of the morning. These include looking at data logs and network traffic. Network traffic is data that moves across a computer network at any point

Infosec analysts may work and meet with teams from all over the world. They update each other on security threats.

in time. They look for anything that is not normal.

If there is a sudden threat alert, infosec analysts brainstorm ways to stop it. They study the threat closely. They might look at ways similar viruses have been prevented.

Infosec analysts help protect hospitals, schools, banks, and governments.

They communicate with other team members, including other infosec analysts. Someone may suggest new software to fix the problem. Then they may try the suggestion and analyze the results. They may repeat this until they find a solution. It may take a long time, or it could be quick. Once the threat is over, infosec analysts write a report. They describe what happened. They discuss how they handled the threat and ways to improve security.

Another task infosec analysts complete is visiting or checking in with users. Analysts answer questions about online security. They may check to see if users are following security guidelines. Infosec analysts often discuss ways to keep computers safe. Work may end in the late afternoon or evening.

This depends on the workplace schedule. Infosec analysts often write to-do lists to prepare for the next day.

WORKING REMOTELY

Some infosec analysts are **remote** employees. Pete Nowicki works as an infosec analyst at a large school. He works remote. He works many miles from the computer systems that he protects. He talks to team members by phone or computer. He likes being able to work from anywhere. He says, "A remote job is hands down the best thing that ever happened to [my] career."[6] Not all infosec jobs can be done remotely. But some companies are willing to hire remote infosec analysts.

These companies provide flexible options for workers.

More companies are looking for remote infosec analysts. These companies provide flexible options for workers.

Working remotely requires a stable internet connection.

CHAPTER FOUR

THE OUTLOOK FOR INFORMATION SECURITY ANALYSTS

Companies are storing more data than ever before. This has led to a rise in cyberattacks. Hackers want to steal personal information for profit. Employers want to keep their data and systems safe. Hiring infosec analysts can help. But this can be challenging.

Infosec analysts are not being trained fast enough to meet the demand. Many companies are looking to hire

Infosec analysts advise companies on how to keep computers secure.

Infosec analysts may provide a company with rules to maintain levels of security. They may also help train employees on ways to be safe while using the internet.

infosec analysts. Some are big tech companies. Others are universities and small businesses.

JOB GROWTH AND SALARY

The US Bureau of Labor Statistics (BLS) is a federal agency that collects, analyzes, and posts data about the economy. It says that infosec analysts are one of the fastest growing careers in the United States. The BLS says the job market could grow as much as 33 percent by 2033. It predicts about 17,300 job openings for infosec analysts each year. There is also a demand for infosec analysts around the world. In 2024, Forbes reported that there were millions of infosec analyst job openings around the world.

In 2024, entry-level analysts made an average of $103,000 per year. Most military analysts make less money. In 2024, they earned between $63,000 and $98,000 per year.

NEW TECHNOLOGY

With new technologies, infosec analysts must stay updated and learn new skills.

AVERAGE SALARY FOR INFOSEC ANALYST BY SPECIALTY

Job Specialty	Average Salary
Cybersecurity Manager	$153,000
Cryptographer Engineer	$137,800
Ethical Hacker	$113,500
IT Auditor	$111,210
Digital Forensic Examiner	$91,030

Source: "10 Cybersecurity Jobs to Know: Entry-Level and Beyond," Coursera, *March 12, 2025. www.coursera.org.*

The average income an infosec analyst earns depends on where they work. This table shows the average income earned in different technological industries.

Hackers will use new technology to commit crimes. But infosec analysts can use new tools to protect user software and programs, too.

Hackers or scammers may use social media to pretend to be people they are not. They may ask for money or private information.

Artificial intelligence (AI) is technology that can perform different tasks and problem solve like humans do. Infosec analysts use AI to accurately spot threats or viruses as soon as they appear. AI detects threats by recognizing any unusual changes in data patterns. AI can then flag the threats for infosec analysts or solve the problems itself. AI can solve these problems through machine learning. Machine learning allows computers to learn from data. It makes predictions to solve problems. It can also protect firewalls that have been breached

Internet of Things

Smart devices play a large role in modern life. Popular smart devices include speakers, TVs, and even refrigerators. These devices create a network known as the Internet of Things (IoT). But many IoT devices are not safe. Hackers use them to get into other systems. One report says, "While the security of IT hardware and software has strengthened in recent years, the security of Internet of Things . . . has not kept pace."

Elizabeth MacBride, "The Dark Web's Criminal Minds See Internet of Things as Next Big Hacking Prize," CNBC, *January 9, 2023. www.cnbc.com.*

by hackers. It can recognize fake emails and bad links, too.

Zero Trust Architecture is a useful framework for users with private information. It assumes that no one can be trusted. Only those who pass several tests can use a computer system. Users must prove their identity in different ways. They may have to give passwords and fingerprints. This may help prevent hackers from accessing personal information.

As technology changes, different roles will be created. New specialties may emerge. For example, cloud security is becoming more important. Many companies are moving to cloud-based services. These programs allow users to store files and information across different internet devices. Users do not need a hard drive to access their information. Cybersecurity professionals must develop skills in cloud security to protect private information. They must also form good relations with cloud service providers.

A career as an infosec analyst offers many advantages. Infosec analysts can work for many kinds of companies. They also have many types of specialties. And their chances of finding jobs are very high. As technology continues to change, infosec analysts will be needed to make sure software programs are up-to-date and protected.

Marc Zirillo has a master's degree in cyber and information security. "Technology is an ever-changing field; we fix things

Infosec analysts engage with both technology and people daily. They ensure users are educated about technology changes.

one day for it to break the next. It speaks strongly to those who enjoy a challenge," he says. "Constantly learning is part of the job, and just as cyber threats evolve, so must the defenders of data."[7]

GLOSSARY

entry-level

at the lowest level of a company or career

firewalls

computer security systems that limit traffic into or out of a private computer network

hard drives

devices that help store data on a computer

malware

short for "malicious software," a kind of virus that tries to harm computers

pop-up windows

windows that suddenly appear on a person's computer screen

recruits

people who just joined a program and are not fully trained

remote

using computer connections to do work from a location other than an office

SOURCE NOTES

INTRODUCTION: FROM GAMER TO INFORMATION SECURITY ANALYST

1. "Anthony Gomez: Information Security Analyst," *Roadtrip Nation*, n.d. https://roadtripnation.com.

CHAPTER ONE: WHAT DOES AN INFORMATION SECURITY ANALYST DO?

2. "Hacked by Jamie Woodruff," *London Speaker Bureau*, n.d. https://us.londonspeakerbureau.com.

CHAPTER TWO: BECOMING AN INFORMATION SECURITY ANALYST

3. Quoted in Madeline Quigley, "Meet Your InfoSec Team: Dean Boenzi, Information Security Analyst III," *Washington University*, September 26, 2024. https://informationsecurity.wustl.edu.

4. "Anthony Gomez: Information Security Analyst."

CHAPTER THREE: A DAY IN THE LIFE OF AN INFORMATION SECURITY ANALYST

5. Casey Lem, "Casey's Answer, What Does Your Typical Day Look Like as an Information Security Analyst?" *CareerVillage,* September 27, 2024. www.careervillage.org.

6. Quoted in Madeline Quigley, "Meet Your InfoSec Team: Pete Nowikow, Information Security Analyst III," *Washington University*, June 27, 2024. https://informationsecurity.wustl.edu.

CHAPTER FOUR: THE OUTLOOK FOR INFORMATION SECURITY ANALYSTS

7. Quoted in "A Day in the Life of a Cybersecurity Professional," *St. John's University*, October 28, 2024. www.stjohns.edu.

FOR FURTHER RESEARCH

BOOKS

Kezia Endsley, *Cybersecurity and Information Security Analysts: A Practical Career Guide.* Rowman & Littlefield, 2021.

Cynthia Kennedy Henzel, *Be a Cybersecurity Specialist.* BrightPoint Press, 2025.

Isabel Teitelbaum, *Be a Computer Support Specialist.* BrightPoint Press, 2025.

INTERNET SOURCES

"Information Security Analyst Overview," *US News & World Report,* November 7, 2024. https://money.usnews.com.

"Information Security Analysts," *US Bureau of Labor Statistics*, August 29, 2024. www.bls.gov.

Shayna Joubert, "Information Security Analysts: Who They Are & What They Do," *Northeastern University,* April 5, 2024. https://graduate.northeastern.edu.

WEBSITES

Air & Space Forces Association's CyberPatriot

www.uscyberpatriot.org

CyberPatriot is a US program for youth. The website advises students on how to become infosec analysts. It also has information about cyber education. It lists camps and competitions.

National Initiative for Cybersecurity Careers and Studies

https://niccs.cisa.gov

The NICCS website offers information for students interested in a cyber security career. It suggests different paths to this career. It has links to internships and apprenticeships. The site also lists clubs, videos, camps, and competitions.

US Army

www.goarmy.com

The US Army website discusses ways to become an information security analyst in the US Army Cyber Corp. It provides the requirements for enlisted soldiers. It also lists benefits. These include education and training, salary, and health care.

INDEX

IMAGE CREDITS

Cover: © DC Studio/Shutterstock Images
5: © Gorodenkoff/Shutterstock Images
7: © Gorodenkoff/Shutterstock Images
9: © Chaikom/Shutterstock Images
10: © Jacob Lund/Shutterstock Images
11: © amgun/Shutterstock Images
13: © Elnur/Shutterstock Images
15: © TippaPatt/Shutterstock Images
16: © Nomad_Soul/Shutterstock Images
19: © Gorodenkoff/Shutterstock Images
20: © Tada Images/Shutterstock Images
22: © Chay_Tee/Shutterstock Images
25: © Bojan Milinkov/Shutterstock Images
26: © fizkes/Shutterstock Images
29: © Jelena Zelen/Shutterstock Images
31: © Gorodenkoff/Shutterstock Images
32: © Studio Romantic/Shutterstock Images
35: © Andrey_Popov/Shutterstock Images
37: © Sfio Cracho/Shutterstock Images
39: © PeopleImages.com-Yuri A./Shutterstock Images
40: © fizkes/Shutterstock Images
43: © Face Stock/Shutterstock Images
44: © PeopleImages.com-Yuri A./Shutterstock Images
47: © Jelena Zelen/Shutterstock Images
49: © Gorodenkoff/Shutterstock Images
50: © PeopleImages.com-Yuri A./Shutterstock Images
52: © Red Line Editorial
53: © tete_escape/Shutterstock Images
57: © PeopleImages.com-Yuri A./Shutterstock Images

ABOUT THE AUTHOR

Miles Herman is a freelance writer living in upstate New York. In his spare time, he loves to research and write about trains and railroading.